Practical Silviculture and Trees Species Management

NIPA® GENX ELECTRONIC RESOURCES & SOLUTIONS P. LTD.
New Delhi-110 034

About the Authors

Dr. Kaushal Singh is a forest ecologist with Masters and Ph.D. in Forestry from Forest Research Institute (Deemed) University, Dehradun, Uttarakhand. This manual is an outcome of his more than a decade of field experience in Silviculture, Ethnobotany, Dendrology (Plant taxonomy), Ecology and biodiversity areas of Madhya Pradesh, Rajasthan, Uttarakhand and Uttar Pradesh. He has published more than 25 research/popular articles in national and international journals and two books. He is a recipient of prestigious S. K. Seth prize-2019. He is currently working as Assistant Professor in Department of Silviculture and Agroforestry, College of Forestry, Banda University of Agriculture and Technology, Banda, Uttar Pradesh.

Dr. Rajiv Umrao is a streamline forester having UG, PG Forestry and Ph.D. in Agroforestry. He has a more than 16 years of Teaching and Research experience. Presently he is working as an Associate Professor, Silviculture and Agroforestry, College of Forestry, Banda University of Agriculture and Technology, Banda Uttar Pradesh, India. He has also served as Assistant Professor at College of Forestry & Environment, Allahabad Agricultural Institute-Deemed University, Allahabad, Uttar Pradesh, India. He has published 35 Research papers in national and international journals and recipient of prestigious *Brandis* and *Young Scientists Award*.

Dr. Sanjeev Kumar is Professor silviculture & Agroforestry at College of Forestry Banda University of Agriculture & Technology Banda, has 16 years of experience of teaching, research and extension in the discipline of forestry. He completed B.Sc. Forestry, M.Sc. and Ph.D. Agroforestry from Dr. Y S parmar University of Horticulture & Forestry, Nauni Solan HP. He started his career in 2008 as Assistant Professor (silviculture & Agroforestry) at college of Horticulture & Forestry Central agricultural University Pasighat Arunachal Pradesh. Presently he is working as Dean college of Forestry Banda University of Agriculture & Technology.

Dr. Garima Gupta completed her PG and Ph.D. in Agroforestry from Bundelkhand University, Jhansi. She has more than a decade of teaching and research experience in the field of Silviculture, carbon sequestration modeling, tree crop interactions, agroforestry soil fertility, and natural resource management. Currently, she is working at Rani Lakshmi Bai Central Agriculture University in Jhansi, Uttar Pradesh. She has authored and published more than 15 research and popular articles in national and international journals. She has been honored with the Excellence in Teaching Award and the Young Achiever Award for her outstanding accomplishments.

Practical Silviculture and Trees Species Management

Kaushal Singh
Assistant Professor
Department of Silviculture and Agroforestry
College of Forestry
Banda University of Agriculture and Technology
Banda-210001, Uttar Pradesh

Rajiv Umrao
Associate Professor
Department of Silviculture and Agroforestry
College of Forestry
Banda University of Agriculture and Technology
Banda-210001, Uttar Pradesh

Sanjeev Kumar
Professor
Silviculture & Agroforestry
College of Forestry
Banda University of Agriculture and Technology
Banda-210001, Uttar Pradesh

Garima Gupta
Department of Silviculture and Agroforestry
College of Horticulture & Forestry
Rani Lakshmi Bai Central Agriculture University
Jhansi-284003, Uttar Pradesh

NIPA® GENX ELECTRONIC RESOURCES & SOLUTIONS P. LTD.
New Delhi-110 034

NIPA® GENX ELECTRONIC
RESOURCES & SOLUTIONS P. LTD.

101,103, Vikas Surya Plaza, CU Block
L.S.C.Market, Pitam Pura, New Delhi-110 034
Ph : +91 11 27341616, 27341717, 27341718
E-mail: newindiapublishingagency@gmail.com
www: www.nipabooks.com

For customer assistance, please contact
Phone: + 91-11-27 34 17 17
Fax: + 91-11-27 34 16 16
E-Mail: feedbacks@nipabooks.com

ISBN: 978-93-58874-95-2

Composed and Designed by NIPA.

Preface

Silviculture, the art and science of forest management, plays a critical role in sustaining and enhancing the health and productivity of our precious forest ecosystems. As professionals, researchers, and students engaged in the field of silviculture, we understand the importance of practical knowledge and hands-on skills in achieving successful forest management outcomes. The authors felt the need to provide updated practical knowledge in the form of a book."

This book has been meticulously designed to provide a comprehensive and practical guide for students with answer of all the practical question of the subject. It covers a wide range of topics, starting with an in-depth exploration of modern silvicultural equipment. We delve into the tools and machinery utilized in the field, providing detailed descriptions and usages. In this manual, we place special emphasis on the morphological description and field identification techniques for various tree species. With detailed illustrations and step-by-step instructions, readers will develop the skills necessary to confidently identify and differentiate between different tree species in the field. Understanding the characteristics and behaviors of tree species is vital for successful silvicultural management. To this end, a comprehensive studies and literature survey on the phenological events exhibited by various tree species has been compiled. By examining their growth patterns, flowering times, and other seasonal events, readers will gain valuable insights into the life cycles and ecological requirements of these trees.

Furthermore, we discuss how tree species respond to environmental factors such as light, fire, drought, frost, and wind. By studying their silvicultural characteristics in relation to these factors, readers will learn how to effectively manage forests in the face of different disturbances. Practical recommendations and management strategies are provided to ensure the resilience and sustainability of tropical tree species in diverse environmental conditions. Silvicultural techniques such as root suckering, coppicing, pollarding, pruning, and browsing have significant implications for the growth and development of tree species. In this book, we present detailed studies on the silvicultural characteristics of tropical and temperate tree species separately in response to these techniques. Through practical demonstrations and step-by-step instructions, readers will acquire the skills necessary to implement these techniques and manipulate tree growth patterns to meet specific management objectives. Seed germination, pollination, and dispersal processes are crucial for the natural regeneration and propagation of tree species. We dedicated an exercise of this manual to the study of these processes, shedding light on the reproductive strategies employed by various tropical and temperate tree species. Practical guidance is provided for enhancing natural regeneration and promoting genetic diversity within forest ecosystems.

The manual also covers specific case studies, including the planting and stand management practices of trees. Additionally, the composition of forest areas and assess the suitability of different tree species for various problematic conditions. By studying the ecological characteristics and requirements of tree species, readers will gain a comprehensive understanding of how to make informed decisions regarding species selection in different ecological contexts.

In conclusion, this practical manual aims to bridge the gap between theory and practice in silviculture and tree species management. It serves as a valuable resource for professionals, researchers, and students seeking practical knowledge and skills in the field. We hope that this manual will empower graduate and postgraduate students of forestry and other allied courses to meet the academic requirement. The authors will greatly appreciate having any errors called to their attention and will welcome comments or criticism from those using the book in teaching or otherwise.

Authors

Contents

Preface .... *v*

1. To Acquaintance with Modern Silvicultural Tools 1
2. To Study the Morphological Description and Field Identification Characteristics of Tropical Tree Species 5
3. To Study the Morphological Description and Field Identification Characteristics of Temperate Tree Species 9
4. To Study the Phenological Events of Tropical Tree Species 13
5. To Study the Phenological Events of Temperate Tree Species 15
6. To Study the Silviculture Characteristics of Trees in Response to Light, Fire, Drought, Frost and Wind of Tropical Tree Species 17
7. To Study the Silviculture Characteristics of Trees in Response to Light, Fire, Drought, Frost and Wind of Temperate Tree Species 19
8. To Study the Silvicultural Characteristics of Trees in Response to Root Sucker, Coppicing, Pollarding, Pruning and Browsing of Tropical Tree Species 21
9. To Study the Seed Germination, Pollination and Dispersal Type of Tropical Tree Species 25
10. To Study the Seed Germination, Pollination and Dispersal Type of Temperate Tree Species 27
11. To Study Tropical Tree Species Managed under Different Silvicultural Systems 29
12. To Study Temperate Tree Species Managed under Different Silvicultural Systems 31
13. To Study the Harmful Insects and Pests of Tropical Tree Species 33
14. To Study the Harmful Insects and Pests of Temperate Tree Species 39
15. To Study the Planting and Stand Management Practices of Teak (*Tectona grandis*) 41

16. To Study the Planting and Stand Management Practices of Kala Shisham (*Dalbergia latifolia*) 45

17. To Study the Planting and Stand Management Practices of Sandal Wood (*Santalum album*) 47

18. To Study the Planting and Stand Management Practices of Poplar (*Populus deltoides*) 49

19. To Study Planting and Stand Management of Bastard Mahogany (*Swietenia macrophylla*) 53

20. To Study Planting and Stand Management of Khair (*Acacia catechu*) 55

21. To Study Planting and Stand Management of Safeda (*Eucalyptus tereticornis*) 59

22. To Study the Species Suitability for Various Problematic Areas 61

23. To Study the Composition of a Forest Area 65

24. To Study the Tending Operation Carried out in a Forest Crop 67

1

To Acquaintance with Modern Silvicultural Tools

Objective: To acquaintance with modern silvicultural tools.

Modern silvicultural tools involve in a paradigm shift from a highly manual to mechanize or digital data recording and tight operational control. Now a days many machines or digital tool are being used in silviculture like pitting machine, tree planter, Laser geo hypsometer, Satellite system etc.

1. Clinometer

Clinometer is a measuring instrument that is used to measure angles (slope), tree height and distance from any object. It works on trigonometric principle and able to measure upward or positive as well as downward or negative slopes.

Fig. 1 : Clinometer

2. Hypsometer

Laser Geo instruments are compact hand-held instrument which offer high precision distance and height measurements. This has a GPS and digital compass that enables data tagging and accurate 3-D measurements such as canopy, map area and heaps of woodchip. Distance measurements can be taken from a few centimeters to over 700m with very high accuracy. Storage (2000 readings) and processing capability is now built in Improved 3 key keypad for easier use.

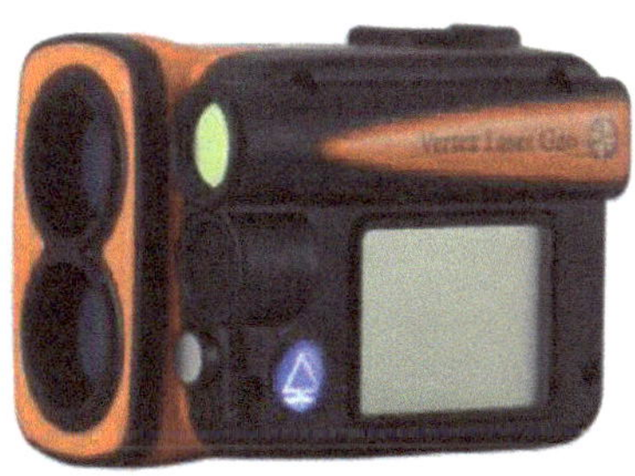

Fig. 2 : Hypsometer

3. Laser Bar Caliper (LBC)

Accurate DBH can be recorded of any desired tree from a distance with the help of Laser caliper which is operated by both hands. Field cost of measurement is reduced with accuracy in measurements of biomass and volume in short time.

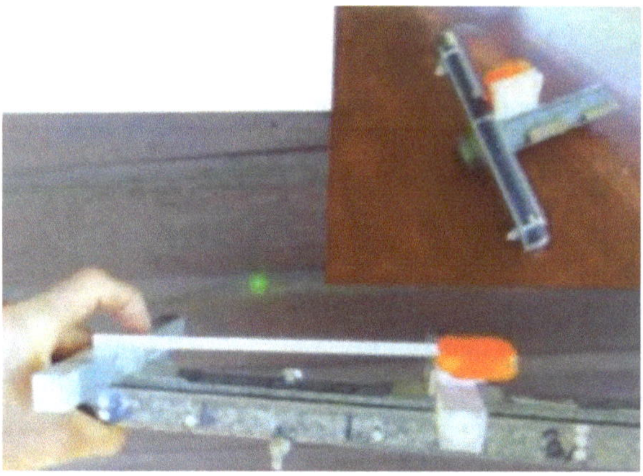

Fig. 3 : Laser Bar Caliper

4. Criterion Laser

A tree height, diameter and range-measuring instrument. The Criterion uses laser light to determine distance from a tree. The user enters numbers read from a rifle-scope and a simple computer chip calculates diameter.

Fig. 4 : Criterion Laser

5. Pitting Machines

Pitting machines is designed to conduct dugout large scale pitting in very short time and uniform in size as compare to manual digging. It is able to dugout approx... 2500 pits per shift. Pits are produced with clear wall without glazing. The sensor of machine has depth sensing capabilities that enables the machine to record only completed pits. Different dimension of pit can be produced by auger.

Fig. 5 : Pitting Machines

6. Tree Planter

Tree planter was launched in Brazil and it is a first of its kind. It is a machine designed to prepare the pit, plant and irrigate the seedlings, therefore called triple planting machine. Tree planter comprises of tracked tractor, carrying three planting heads. Each head has the capacity to plant 196 seedlings before being refilled. The tracked dozer carrier pulls a 5000l tandem axle water tank which supplies water for irrigation during the planting process. It is capable of operating at spacings between 3m or 3.5m between the planting heads. Every planting head can move sideways up to a maximum of 60 cm to improve the positioning of seedlings. It also has an autopilot feature which enables the machine too drive infield using a pre-defined path. Seedling positioning is registered using a Hexagon system, which guides the planting heads without human action. About 900 seedlings/hour planting and irrigating can be planted within 12 seconds of cycle.

Fig. 6 : Tree Planter

7. Satellite system

Satellites such as Advanced Very High-Resolution Radiometer (AVHRR) and Moderate resolution Imaging Spectrometer (MODIS) were launched during 1999. AVHRR and MODIS provides global imagery at the interval of one or two days or twice a day with 1 km resolution for temperature, vegetation cover, deforestation rate, cloud, snow etc.

8. Wireless Sensor Network (WSN)

With the help of the new emerging technology, we can detect the forest fire and save large area before burning. **Wireless Sensor Network (WSN)** is the example of such technology that can alarm the forest fire. The line-of-sight problem of optical cameras in forests can be solved with the second type of sensors. The same printed circuit board integrates the wireless transceiver, sensors and data processing. Sensors are able to influence the parameters like pressure, radiations, temperature, gases, humidity and many more. Sensor network normally deployed in a large-scale random distribution on remote or inaccessible places and under harsh environment for a certain period of lifetime. This technology relied on low data rate and short ranges of communication with multi-hope fashion to reach the sink. Sensor network and IP cameras are used in Spain to detect the fire ignition by sensors and use the closest camera to provide images. Whereas, in South Korea a **Forest Fire Surveillance System (FFSS)** developed. This system done the surveillance work by sensor network to observe temperature, illumination and humidity. Theses observation recorded in a database to make a daily calculation and comparison for any danger.

2

To Study the Morphological Description and Field Identification Characteristics of Tropical Tree Species

Objective: To study the morphological description and field identification characteristics of various tropical tree species.

Observation recording: For correct identification of any tropical tree species, it is necessary to observe the type of leaf, leaf arrangement, surface ornamentation of leaf and various types of margins. The type of leaf, margins and leaf arrangement can be seen without help of any lens whereas to observe the ornamentation a hand lens will be require. Here some examples are given below just for better understanding of the students.

Leaf Arrangement

Alternate: One leaf is found on each node usually form a twisting pattern as in *Holoptelea integrifolia* (Churel), *Pongamia pinnata* (Karanj), *Butea monosperma* (dhak) etc.

Opposite: At each node bears a pair of leaves. The pairs of successive leaves may be parallel (superposed) as in Guava or at right angles (decussate) as in Teak and Madar.

Whorled (verticillate): In whorled phylotaxy 3 or more then 3 leaves found at each node as in *Alstonia, Callicarpa, Raulwolfia* etc.

Leaf surface

Leaf surface characteristics are important in diagnosis of several taxa. Leaf surface may be covered by trichomes arranged variously.

Glabrous: If there is no hairs or any other ornamentation present on surface of lamina as in Semal, Mango, Jamun, Ashok, Mahagony, Toon (Singh *et al.*, 2021) etc.

Glaucous: Surface of leaf blade with waxy coating, which easily rubs off as in genus *Citrus*, *Eucalypus*, *Trichillia connaroides* etc.

Glutinous: Surface of leaf blade coated with sticky exudation as in *Tobacco nicotiana*.

Hispid: Leaf blade is coverered with stiff hairs as in *Arnebia hispidissima*, *Ficus hispida*.

Lanatus: When covered with long, intertwined and wooly hairs as in *Verbascum thapsus*

Pilose: Lamina is covered with straight and sparse hairs as in *Indigofera astragalina*, lower surface of *Dalbergia lanceolaria* etc.

Pubescent: Leafblad is covered with straight, soft and short hairs as in *Flacourtia indica*, *Terminalia coronata*, *Cocculus hirsutus*, *Averrhoa carambola*, *Kydia calycina, Indigofera hamiltonii* etc.

Rugose: Surface have reticulated lines with convex space between the line as in *Plecteranthus rugosu* (Upper surface), *Euonymus tingens* etc.

Scabrous: Surface of lamina feels tough due to short rough points as in *Streblus asper, Helecteris isora*, *Tectona grandis*, *Nyctanthes arbor tristis* etc.

Sericeous: Leaf blade is covered with fine, soft and silky hairs as in *Desmodium brachystachyum* (Lower surfacea), *Anogeissus acuminata* (Lower surfacea),

Tomentose: Covered with densely matted soft hairs, woolly in appearance as in *Terminalia elliptica*, *Miliusa velutina*, *Cochlospermum religiosum* etc.

For clarification of the students some examples are given in below table:

S. No.	**Tree Species**		**Identification characteristics**			
	Scientific Name	**Vernacular Name**	**Phyllo-taxy**	**Leaf type**	**Leaf surface**	**Leaf margin**
1	*Acacia catechu*	Khair	Alternate	Bipinnate	Pubescent	Entire
3	*Acacia nilotica*	Desi babool	Alternate	Bipinnate	Pubescent	Entire
4	*Ailanthus triphysa*	Malabar tree of heaven	Alternate	Imparipinnate	Glabrous	Entire
5	*Albizia lebbeck*	Kala Siris	Alternate	Bipinnate	Glabrous	Entire
6	*Albizia procera*	Safed Siris	Alternate	Bipinnate	Smooth and glossy	Entire
7	*Anogeissus latifolia*	Bakli	Alternate	Simple	Glossy	Entire Smooth
8	*Anogeissus pendula*	Dhok	Alternate	Simple	Glossy	Entire

9	*Azadirachta indica*	Neem	Opposite	Imparipinnate	Smooth shinny	serrated
10	*Casuarina equisetifolia*	Casuarina	Whorl	Needle shaped	Glabrous	Entire
12	*Dalbergia latifolia*	Rosewood	Opposite	Simple	Smooth and slightly glossy	Entire
13	*Dalbergia sissoo*	Shisham	Opposite	Simple	Smooth and glossy	Entire
14	*Eucalyptus tereticornis*	Safeda	Alternate/ Opposite	Simple	glamorous	Entire
15	*Grevillea robusta*	Silver oak	Alternate	Bipinnate	Smooth and glossy	serrated
16	*Hopea parviflora*		Alternate	Simple	Hairy	Entire
17	*Lagerstroemia microcarpa*	Ben teak	Opposite	Simple	Glamourous	Entire
18	*Leucaena leucocephala*	Su-babool	Alternate	Bipinnate	Smooth, shiny	Entire
19	*Melia dubia*	Malabar Neem	Opposite	Imparipinnate	Smooth	serrated
20	*Pongamia pinnata*	Karanj	Alternate	Imparipinnate	Glabrous	Entire
21	*Pterocarpus marsupium*	Bijasar	Alternate	Imparipinnate	Glabrous	Entire smooth
22	*Pterocarpus santalinus*	Rakt chandan	Alternate	Imparipinnate	Glabrous	Entire smooth
23	*Santalum album*	Chandan	Opposite	Simple	Shiny smooth& dark green	Entire smooth
24	*Shorea robusta*	Sal	Alternate, Spiral	Simple, oblong, ellipate	Dark green	Entire smooth
25	*Swietenia macrophylla*	Mahogany	Alternate	Simple	Smooth glossy shiny apper	Entire smooth
26	*Tectona grandis*	Teak, Sagon	Alternate	Simple, broad	Glossy & dark green	Entire smooth
27	*Terminalia arjuna*	Arjun	Alternate	Simple	Smooth	Entire
28	*Terminalia bellerica*	Bahera	Alternate	Simple	Glabrous	Entire
29	*Terminalia tomentosa*	Sain, Sadar	Opposite	Simple	Glabrous	Entire

3

To Study the Morphological Description and Field Identification Characteristics of Temperate Tree Species

Objective: To study the morphological description and field identification characteristics of various temperate tree species.

Observation recording: For correct identification of any temperate tree species, it is necessary to observe the type of bark, type of leaf, leaf arrangement and surface of leaf. For easy understanding of the students some examples are given in table:

S. No.	Tree Species	Identification characteristics					
	Scientific Name	**Vernacular Name**	**Bark**	**Leaf type**	**Phyllotaxy**	**Leaf surface**	**Leaf margin**
1	*Abies pindrow*	Fir	Grey brown longitudinally fissured	Needle Shape flattened	Pectinate	Dark green shiny	Entire
2	*Acer oblongum*	Himalayan maple	Brownish grey reticulately fissured	Simple	Opposite	Glabrous above and grey beneath	Entire
3	*Betula utilis*	Bhoj patra	White or brownish peels off in very thin horizontal rolls	Ovate-lanceolate	Alternate	Pubescent in young and gland dotted in older	Serrate
4	*Cedrus deodara*	Deodar	Greyish-brown with vertical and diagonal cracks	Needle Shape (15-30 in cluster)	Spirally arranged	Glaucous or silvery	Entire
5	*Cupressus torulosa*	Bhutan Cypress	Grey-brown peeling in long thin stripes	Scale like	Opposite	Glaucous	Entire
6	*Juglans regia*	Walnut	Light-grey with longitudinal fissures	Imparipinnate	Opposite	Thickly tomentose	Entire
7	*Juniperus macropoda*	Pencil Cedar	Reddish brown thick exfoliate in long fibrous stripes	Scale like	Opposite	Glaucous	Entire
8	*Picea smithiana*	Spruce	Greyish-brown, shallow racked	Acicular	Spirally arranged	Glabrous	Entire
9	*Pinus gerardiana*	Chilgoza	Silver grey, thin flaking in plates	Needle Shape	Spirally arranged	Glossy green	Entire

10	*Pinus roxburghii*	Chir	Slate-grey, smmoth but later on become rough and shallow fissured	Needle Shape (3 in a cluster)	Spirally arranged	Glaucous and glabrous	Entire
11	*Pinus wallichiana*	Kail	Slate-grey, smooth but later on become rough and shallow fissured	Needle Shape (5 in a cluster)	Spirally arranged	Glaucous and glabrous (Dark bluish-green)	Entire or minutely toothed
12	*Quercus floribunda (2201-2700m)*	Moru Oak	Dark grey or black, peeling in longitudinal scales	Lanceolate	Alternate	Glabrous	Entire or spinous-serrate
13	*Quercus leucotrichophora (1500-2200m)*	Banj Oak	Dark grey rough with crcks and fissures	Obate-lnceolate	Alternate	Glabrous	Toothed
14	Quercus semecarpifolia (2701-3300m)	Kharsu Oak	Drak grey, cracked more or less four sided scales	Ellepticor oblong	Alternate	Glabraos and dark green above and rusty tomentos beneath	Toothed or Spiny in young tree and entire in old trees
15	*Rhododendron arboreum*	Burans	Reddish brown, peeling in small flakes	Elleptic-oblong	Alternate	Globrous above and rusty tomentose beneath	Entire
16	*Taxux baccata*	Common Yew	Reddish grey, thin and scaly	Linear	Alternate	Brownish yellow and mealy beneath	Entire

4

To Study the Phenological Events of Tropical Tree Species

Objective: To study the phenological events of various tropical tree species.

Observation recording: Phenological events are typically quantified by observing the date of onset and the duration of particular phenophases, which include both vegetative and reproductive events like leaf shedding, leaf emergence, flowering and fruiting. To record the phenological events students should note down the emergence of leaf, flowering and fruiting at the interval of 15 days by vising the targeted tree species. For easy understanding of the students some examples of tropical trees are given in below table:

S. No.	Scientific Name	Phenological events			
		Leaf shedding	**Leaf emergence**	**Flowering**	**Fruiting**
1	*Acacia catechu*	Feb-March	April-May	April/May-July/August	Sept-Oct.
2	*Acacia leucophloea*	Feb-March	April	July -Nov.	April-June
3	*Acacia nilotica*	March-June	March-June	April-June	Feb- March
4	*Ailanthus triphysa*	Evergreen		Feb-March	March-April
5	*Albizia lebbeck*	Oct-Nov	March/April	Mach-May	Dec-Nov.
6	*Albizia procera*	April-June	April-May	June- august	Feb-April
7	*Anogeissus latifolia*	Feb-April	Feb-April	June-Sept	Dec-March
8	*Anogeissus pendula*	March-April	April-June	August-Sept	Dec-Jan
9	*Azadirachta indica*	Late Feb-March	Early April	Mid-April	July
10	*Casuarina equisetifolia*	Evergreen		Feb-April/ Sept/Oct	June-Dec
12	*Dalbergia latifolia*	Jan-Feb	Late April-May	April-June	Dec- Jan
13	*Dalbergia sissoo*	Nov-Dec	Jan- Feb	March-April	Nov- Jan
14	*Eucalyptus tereticornis*	Evergreen		May-June/ Oct-Nov	July/Dec

15	*Grevillea robusta*	Jan-Feb	End of March	March- mid April	May-June
16	*Hopea parviflora*	After emergence	Dec//Jan-April/may	Jan- May	April- June
17	*Lagerstroemia microcarpa*	Feb- April	May- June	Mach -April	May- June
18	*Leucaena leucocephala*	Dec-Jan	Feb	Feb-July	April-June
19	*Melia dubia*	Dec- Feb	Feb-March	Feb – March	Oct- Feb
20	*Pongamia pinnata*	March- April	Late April	April-May	April -May
21	*Pterocarpus marsupium*	Mid to late March	Early May	Late Sept	Feb-March
22	*Pterocarpus santalinus*	Feb -April/ May	Jan- March	April-July	Sept-Nov
23	*Santalum album*	Depending the on host nature if the host is evergreen behaves like evergreen and vise-versa		Dec-March	Nov-Dec
24	*Shorea robusta*	Feb -mar	Late Feb -April	March -April	June-July
25	*Swietenia macrophylla*	Feb -march	March	April – May	Feb- March
26	*Tectona grandis*	Nov	April-June	July- August	Nov- Jan
27	*Terminalia arjuna*	March-April	Late March	April-May	March-May
28	*Terminalia bellerica*	Nov-Jan	April-May	April-June	Nov-Feb
29	*Terminalia tomentosa*	Jan -March	End of June	July	Oct.

5

To Study the Phenological Events of Temperate Tree Species

Objective: To study the phenological events of various temperate tree species.

Observation recording: Phenological events are typically quantified by observing the date of onset and the duration of particular phenophases, which include both vegetative and reproductive events like leaf shedding, leaf emergence, flowering and fruiting. To record the phenological events students should note down the emergence of leaf, flowering and fruiting at the interval of 15 days by vising the targeted tree species. For easy understanding of the students some examples of temperate trees are given in below table:

S. No.	Tree Species	Phenological events				
	Scientific Name	Leaf shading	Leaf emergence	Flowering	Fruiting	Seed maturity
1	*Abies pindrow*	May-June	April-May	April	May	Sept-Oct.
2	*Acer caesium*	Nov.	March-April	May-June	June-July	Oct-Nov.
3	*Betula utilis*	Nov.	Feb-March	March-April	May-June	Oct-Nov.
4	*Cedrus deodara*	May	March	June	Sept.	Sept-Nov.
5	*Cupressus torulosa*	April	April-May	Sept-October	Feb	April (After 12 months)
6	*Juglans regia*	Oct-Nov.	March-April	Mid-April	April-May	Sept. Oct.
7	*Juniperus macropoda*	Autumn	Spring	Feb.-March		Sept.-October
8	*Picea smithiana*	May-June	April	April-May	July	October-Nov.
9	*Pinus gerardiana*	May-June	April	May-June	July	Sept-Oct.
10	*Pinus roxburghii*	May-June	January	Jan (male Cone), Feb (Female Cone)	Feb.	March (after 24 Months)
11	*Pinus wallichiana*	June-July	March-April	April	May	Sept-Oct.

12	*Quercus floribunda (2201-2700m)*	March-May	April-May	May-June	May-July	June-July (after 12 month)
13	*Quercus leucotrichophora (1500-2200m)*	March-May	March-April	April-May	May-July	Nov-Dec. (after 15 month)
14	*Quercus semecarpifolia (2701-3300m)*	March-May	April-May	May-June	May-July	Aug - Sept. (After 15 months)
15	*Rhododendron arboreum*	March-April	April-May	March-May	April-May	Oct. Dec.
16	*Taxux baccata*	Evergreen		March-May	Aug-Nov.	Aug-Nov.

6

To Study the Silviculture Characteristics of Trees in Response to Light, Fire, Drought, Frost and Wind of Tropical Tree Species

Objective: To study the silviculture characteristics of trees in response to light, fire, drought, frost and wind of various tropical tree species.

Observation recording: To record the response of various tree species to the light, fire, drought, frost and wind, frequent visits of the forest areas which are affected by fire, drought and frost should be conducted. Studies relevant to light requirement in various growth stages of tree and ability to stand against various intensities wind are very lengthy that requires long time periodic studies. Here some examples of various tree responses are given in below table for better understanding of the students:

S. No.	Tree Species		Tree response			
	Scientific Name	**Vernacular Name**	**Light**	**Fire**	**Drought**	**Frost**
1	*Acacia catechu*	Khair	Strong light demander	Moderately Resistant	Hardy	Hardy
2	*Acacia leucoploea*	Remja	Shade demander		Resistant	Tender
3	*Acacia nilotica*	Desi babool	Shade demander	Tolerant	Resistant	Tender
5	*Albizia lebbeck*	Kala Siris	Light deander		Hardy	Hardy
6	*Albizia procera*	Safed Siris	Light deander		Hardy	Hardy
7	*Anogeissus latifolia*	Bakli	Shade demander	Tolerant	Resistant	Resistant
8	*Anogeissus pendula*	Dhok	Light deander		Resistant	Hardy
9	*Azadirachta indica*	Neem	Light deander	Tender	Hardy	Sensitive

10	*Casuarina equisetifolia*	Casuarina	Shade demander		Hardy	
11	*Dalbergia latifolia*	Rosewood	Moderately demander	Sensitive	Resistant	Tender
12	*Dalbergia sissoo*	Shisham	Light demander	Moderately Resistant	Hardy	Hardy
13	*Eucalyptus tereticornis*	Safeda	Shade demander	Moderately Resistant	Resistant	Resistant
14	*Grevillea robusta*	Silver oak	Light demander			Resistant
15	*Lagerstroemia microcarpa*	Ben Teak		Resistant	Resistant	Tender
16	*Leucaena leucocephala*	Subabool	Moderately light demander	Resistant	Resistant	Tender
17	*Pterocarpus marsupium*	Bijasal	Moderately light demander	Hardy	Hardy	Tender
18	*Pterocarpus santalinus*	Rakt Chandan	Light demander	Resistant	Resistant	
19	*Shorea robusta*	Sal	Moderately light	Resistant	Sensitive	
20	*Swietenia macrophylla*	Mahogni	Light demander	Sensitive	Sensitive	Tender
21	*Tectona grandis*	Teak, Sagon	Light demander	Sensitive	Sensitive	Tender
22	*Terminalia arjuna*	Arjun	Moderate shade bearer		Sensitive	sensitive
23	*Terminalia bellerica*	Bahera	Light demander		Hardy	Sensitive

7

To Study the Silviculture Characteristics of Trees in Response to Light, Fire, Drought, Frost and Wind of Temperate Tree Species

Objective: To study the silviculture characteristics of trees in response to light, fire, drought, frost and wind of various temperate tree species.

Observation recording: To record the response of various tree species to the light, fire, drought, frost and wind, frequent visits of the forest areas which are affected by fire, drought and frost should be conducted. Studies relevant to light requirement in various growth stages of tree and ability to stand against various intensities wind are very lengthy that requires long time periodic studies. Here some examples of various tree responses (Troup, 1921) are given in below table for better understanding:

S. No.	Tree Species	Tree response					
	Scientific Name	**Verna-cular Name**	**Light**	**Fire**	**Drought**	**Frost**	**Wind**
1	*Abies pindrow*	Silver Fir	Shade bearer (seedling and Sapling stage)	Sensitive	Sensitive	Sensitive	Moder-ate wind firm
2	*Acer caesium*	Himalayan maple	demander	Sensitive	Hardy	Hardy	Moder-ate wind firm
3	*Betula alnoides*	Bhoj Patra	demander	Resistant	Hardy	Hardy	firm
4	*Cedrus deodara*	Deodar	Shade bearer (seedling stage)	Tender	Sensitive	Hardy	Firm
5	*Cupressus torulosa*	Himalayan Cypress	Shade bearer (seedling stage)	Tender	Resistant	Hardy	Firm

6	*Juglans regia*	Walnut	Demander	Sensitive	Sensitive	Sensitive	Firm
7	*Juniperus macropoda*	Pencil Cedar	Demander	Sensitive	Hardy	Hardy	Firm
8	*Picea smithiana*	Spruce	Moderate demander	Tender	Resistant	Hardy	Thrown
9	*Pinus gerardiana*	Chilgoza	Demander	Sensitive	Resistant	Hardy	Firm
10	*Pinus roxburghii*	Chir	Demander	Resistant	Resistant	Hardy	Firm
11	*Pinus wallichiana*	Kail	Strong light demander	Sensitive	Resistant	Hardy	Firm
12	*Quercus floribunda*	Moru Oak	Demander	Sensitive	Sensitive	Hardy	Firm
13	*Quercus leucotri-chophora*	Banj Oak	Moderate light demander	Sensitive	Sensitive	Resistant	Firm
14	*Quercus seme-carpifolia*	Kharsu Oak	Demander	Sensitive	Sensitive	Resistant	Firm
	Salix alba	Willow	Strong light demander	Sensitive	Sensitive	Hardy	Firm
15	*Taxus baccata*	Common Yew	Shade bearer	Sensitive	Tolerant	Tolerant	Firm

8

To Study the Silvicultural Characteristics of Trees in Response to Root Sucker, Coppicing, Pollarding Pruning and Browsing of Tropical Tree Species

Objective: To study the silvicultural characteristics of trees in response to root sucker, coppicing, pollarding, pruning and browsing of various tropical tree species

Observation recording: To record the response of various tree species to produce root sucker, coppicing, pollarding, pruning and browsing of various tropical tree species requires field visits and conducting experiment to observe which species is producing root suckers, coppice etc. Here some examples of various tree responses are given in below table for understanding of the students:

S. No.	Tree Species		Tree response				
	Scientific Name	**Vernacular Name**	**Root sucker**	**Coppicing**	**Pollarding**	**Pruning**	**Browsing**
1	*Acacia catechu*	Khair	fairly	Weak coppicer	Fair	occasionally	Seedlings browsed
2	*Acacia leucophloea*	Ronj, Remja	fairly	poor	Fair	occasionally	Seedlings browsed
3	*Acacia nilotica*	Desi babool	No	poor	Well	Freely lopped	Seedlings browsed
4	*Ailanthus excelsa*	Malabar tree of heaven	no	well	Good	fodder	Seedlings browsed
5	*Albizia lebbeck*	Kala Siris	fairly	well	Good	fodder	Seedlings browsed
6	*Albizia procera*	Safed Siris	fairly	well	Good	fodder	Seedlings browsed
7	*Anogeissus latifolia*	Bakli	fairly	well	Well	fodder	Seedlings browsed
8	*Anogeissus pendula*	Dhok	fairly	well	Well	fodder	Seedlings browsed
9	*Azadirachta indica*	Neem	Good	well	Well	Frequent	susceptible
10	*Butea monosperma*	Flame of forest, palash	good	well	Well	Frequent	susceptible
11	*Casuarina equisetifolia*	Casuarina	no	poor	No	Frequent	Not browsed
12	*Dalbergia latifolia*	Rosewood	good	well	Well	Frequent	susceptible
13	*Dalbergia sissoo*	Shisham	good	strong	No	Frequent	susceptible
14	*Diospyros melanozylon*	Tendu	good	moderate	Good	Freely lopped	Not browsed
15	*Eucalyptus tereticornis*	Safeda	No	good	Fair	Frequent	termite
16	*Hardwickia binata*	Anjan	fairly	well	Good	fodder	susceptible
17	*Leucaena leucocephala*	Subabool	good	good	Good	Frequent	susceptible
18	*Madhuca longifolia*	Mahua	no	moderate	No	no	susceptible
19	*Melia dubia*	Malabar Neem	no	well	Good	fodder	susceptible
20	*Pongamia pinnata*	Karanj	good	well	Fair	Frequent	susceptible
21	*Populus deltoides*	poplar	No	Moderate	Good	Frequent	susceptible
22	*Pterocarpus marsupium*	Bijasar	fair	good	Good	Frequent	susceptible
23	*Pterocarpus santalinus*	Rakt Chandan	fair	excellent	Good	Frequent	susceptible

24	*Santalum album*	Chandan	fair	well	Good	Frequent	susceptible
25	*Shorea robusta*	Sal	poor	well	No	Frequent	susceptible
26	*Swietenia mahogani*	Mahogani	no	good	Well	Frequent	susceptible
27	*Tectona grandis*	Teak, Sagon	poor	moderate	Well	Frequent	Not browsed
28	*Terminalia arjuna*	Arjun	good	good	Well	Frequent	susceptible
29	*Terminalia bellerica*	Bahera	good	good	Well	Frequent	susceptible
30	*Terminalia tomentosa*	Sain, Sadar	good	well	Well	Frequent	susceptible

9

To Study the Seed Germination Pollination and Dispersal Type of Tropical Tree Species

Objective: To study the seed germination, pollination and dispersal type of various tropical tree species.

Observation recording: To record the observation on seed germination, Pollination and dispersal needs to conduct frequent field visits in different seasons for recording the various pollination and dispersing agents and conducting experiment on seed germination. Some examples are given in below table for understanding of the students:

S. No.	Scientific Name	Seed Germination	Seed dispersal	Pollination
1	*Acacia catechu*	Epigeal	Hydrochory/ Anemochory	Hydrophily/ Anemophily
2	*Acacia leucophloea*	Epigeal	Zoochory	Zoophily
3	*Acacia nilotica*	Epigeal	Hydrochory/ Endozoochory	Hydrophily/ Endozoophily
4	*Ailanthus excelsa*	Epigeal	Chiropterochory/ Ornithochory	Chiropterophily/ Ornithophily
5	*Albizia lebbeck*	Epigeal	Anemochory	Anemophily
6	*Albizia procera*	Epigeal	Endozoochory	Endozoophily
7	*Anogeissus latifolia*	Epigeal	Anemochory	Anemophily
8	*Anogeissus pendula*	Epigeal	Anemochory	Anemophily
9	*Azadirachta indica*	Epigeal	Ornithochory/ Chiropterochory	Ornithophily/ Chiropterophily
10	*Butea monosperma*	Hypogeal	Anemochory	Anemophily
11	*Casuarina equisetifolia*	Epigeal	Hydrochory	Hydrophily
12	*Dalbergia latifolia*	Epigeal	Anemochory	Anemophily
13	*Dalbergia sissoo*	Epigeal	Anemochory	Anemophily
14	*Diospyros melanozylon*	Epigeal	Ornithochory / zoochory	Ornithophily / zoophily

15	*Eucalyptus tereticornis*	Epigeal	Anemochory	Anemophily
16	*Hardwickia binata*	Epigeal	Anemochory	Anemophily
17	*Leucaena leucocephala*	Epigeal	Zoochory	Zoophily
18	*Madhuca longifolia*	Hypogeal	Ornithochory / Entomochory	Ornithochory / Entomophily
19	*Melia dubia*	Epigeal	Anemochory	Entomophily
20	*Pongamia pinnata*	Hypogeal	Hydrochory	Hydrophily
21	*Populus deltoides*	Epigeal	Anemochory	Anemophily
22	*Pterocarpus marsupium*	Epigeal	Anemochory	Anemophily
23	*Pterocarpus santalinus*	Epigeal	Anemochory	Anemophily
24	*Santalum album*	Epigeal	Ornithochory	Ornithophily
25	*Shorea robusta*	Hypogeal	Anemochory	Anemophily
26	*Swietenia mahogani*	Hypogeal	Anemochory	Anemophily
27	*Tectona grandis*	Epigeal	Hydrochory	Hydrophily
28	*Terminalia arjuna*	Hypogeal	Anemochory/ Hydrochory	Anemophily
29	*Terminalia bellerica*	Hypogeal	Anemochory	Anemophily
30	*Terminalia tomentosa*	Epigeal	Hydrochory/ Anemochory	Anemophily

10

To Study the Seed Germination Pollination and Dispersal Type of Temperate Tree Species

Objective: To study the seed germination, pollination and dispersal type of various temperate tree species.

Observation recording: To record the observation on seed germination, Pollination and dispersal needs to conduct frequent field visits in different seasons for recording the various pollination and dispersing agents and conducting experiment on seed germination. Some examples of temperate tree species are given in below table for understanding of the students:

S. No.	Scientific Name	Vernacular Name	Type of Seed Germination	Type of Seed Dispersal	Pollination
1	*Abies pindrow*	Silver Fir	Epigeal	Anemochory	Anemophily
2	*Acer caesium*	Himalayan maple	Epigeal	Anemochory	Anemophily
3	*Betula alnoides*	Bhoj Patra	Hypogeal	Anemochory	Anemophily
4	*Cedrus deodara*	Deodar	Epigeal	Anemochory	Anemophily
5	*Cupressus torulosa*	Himalayan Cypress	Epigeal	Anemochory	Anemophily
6	*Juglans regia*	Walnut	Hypogeal	Barochory	Barophily
7	*Juniperus macropoda*	Pencil Cedar	Epigeal	Anemochory	Anemophily
8	*Picea smithiana*	Spruce	Epigeal	Anemochory	Anemophily
9	*Pinus gerardiana*	Chilgoza	Epigeal	Anemochory	Anemophily
10	*Pinus roxburghii*	Chir	Epigeal	Anemochory	Anemophily
11	*Pinus wallichiana*	Kail	Epigeal	Anemochory	Anemophily
12	*Quercus floribunda*	Moru Oak	Hypogeal	Barochory	Barophily

13	*Quercus leucotrichophora*	Banj Oak	Hypogeal	Barochory	Barophily
14	*Quercus semecarpifolia*	Kharsu Oak	Hypogeal	Barochory	Barophily
15	*Salix alba*	Willow	Epigeal	Anemochory	Anemophily
16	*Taxus baccata*	Common Yew	Epigeal	Anemochory	Anemophily

11

To Study Tropical Tree Species Managed under Different Silvicultural Systems

Objective: To study various tropical tree species managed under different silvicultural systems.

Observation recording: Today green felling is completely banned. Fuelwood species are managed under clear-felling systems and tree species infected by insects and pests are felled selectively. The students are advised to visit the fuelwood plantations and forests areas where selective felling is going on. Some examples of tropical tree species are given in below table for help of the students:

Tree species	Silviculture system
Acacia catechu	Clear Felling (Shiwalik, Uttarakhand), Coppice selection (Himachal Pradesh)
Acacia leucophloea	Clear Felling
Acacia nilotica	Clear Felling
Ailanthus triphysa	Clear Felling
Albizia lebbeck	Clear Felling
Albizia procera	Clear Felling
Anogeissus pendula	Coppice with standard (Rajasthan)
Azadirachta indica	Clear Felling
Casuarina equisetifolia	Clear Felling (Coastal areas)
Dalbergia latifolia	Shelterwood, Selection
Dalbergia sissoo	Clear Felling
Eucalyptus tereticornis	Clear Felling (U.P), Simple coppice (U.P, Haryana, Tamil Nadu),
Hopea parviflora	Selection, Irregular Shelterwood, Selection cum Improvement Felling (South India)
Melia dubia	Clear Felling
Populus deltoides	Clear Felling
Pterocarpus marsupium	Shelterwood, Selection

Pterocarpus santalinus	Shelterwood, Selection
Santalum album	Selection (Tamil Nadu)
Shorea robusta	Selection (UK, M.P, Chattisgarh), Clear felling (U.P), Alternate Strip, Shelterwood (Orissa, Assam, U.P, M.P) Coppice with reserve (M.P), Selection Coppice (Orissa), Simple Coppice (U.P, M.P, Bihar, Orissa, Assam, Tripura, West Bengal), Coppice with standard (U.P, Bihar, Orissa)
Tectona grandis	Clear Felling (U.P, Uttarakhand, Maharashtra, West Bengal, Kerala), Selection (M.P)
Terminalia arjuna	Clear Felling
Terminalia bellerica	Clear Felling
Terminalia tomentosa	Clear Felling

12

To Study Temperate Tree Species Managed under Different Silvicultural Systems

Objective: To study various temperate tree species managed under different silvicultural systems.

Observation recording: Now a days green felling is completely banned. Fuelwood species are managed under clear-felling systems and tree species infected by insects and pests are felled selectively. The students are advised to visit the fuelwood plantations and forests areas where selective felling is going on. Some examples of temperate tree species are given in below table for help of the students:

Tree Species	Silviculture System used
Abies pindrow	Clear Felling, Selection (Himachal Pradesh, Uttarakhand)
Acer caesium	Shelterwood, Selection
Cedrus deodara	Uniform (Uttarakhand) Group (Himachal Pradesh, Uttarakhand), Indian Irregular Shelterwood (Himachal Pradesh, Uttarakhand)
Cupressus torulosa	Clear Felling (West Bengal)
Picea smithiana	Clear Felling, Selection (Himachal Pradesh, Uttarakhand), Shelterwood
Pinus gerardiana	Clear Felling
Pinus roxburghii	Clear Felling
Pinus wallichiana	Clear Felling
Quercus floribunda	Shelterwood, Selection
Quercus leucotrichophora	Shelterwood, Selection
Quercus semecarpifolia	Shelterwood, Selection
Rhododendron arboreum	Selection
Taxux baccata	Selection

13

To Study the Harmful Insects and Pests of Tropical Tree Species

Objective: To study the harmful insects and pests of various tropical tree species.

Observation recording: The students are advised to visit the different forests areas and identify various insects and pests damaging trees in various ways like leaf feeding (defoliator), stem boring (stem borer) damaging seeds (seed borer) etc and various pathogens. Some examples of tropical tree species are given in below table:

S. No.	Scientific Name	Insects	Disease and pathogen
1	*Acacia catechu*	Leaf eater (*Dasychira mendosa*) Seed borer (*Bruchidus terranus*)	Heart rot (*Fomes badius*) Root rot (*Ganoderma lucidum*) Powdery mildew (*Erysiphe acacia*)
2	*Acacia leucophloea*	Seeds attacker (*Bruchidius andrewesi*)	Gall rust (*Hapalophragmium penderosum*)
3	*Acacia nilotica*	Root borer beetle (*Celosterna scabrator*) Defoliater beetle (*Psiloptera fastusa*) Defoliater larva (*Ascotis infixaria, Cusiala raptaria, Hyposidra successaria, Semiothesa streniataria disputaria*) Pod borer larva (*Argyroploce illepida*) Bark feeding and scorched leaf (*Pteroma plagiophleps*) Damage seedlings and saplings (*Hypolixus truncatulus*)	Root rot (*Ganoderma lucidun*) Spongy rot (*Fomes badius*)
4	*Ailanthus triphysa*	Whole plant damage (*Hemiptera Reduviidae*)	Shot-hole (*Glomerella cingulata*) Root rot (*Rhizoctonia solani*)

5	*Albizia lebbeck*	Defoliator (*Bamra mundata, Enmonodia vespertilo, Polydesma umbricola, Rhesala imparata, Speiredonia retorta, Eriboesa alhamas and Eurema blanda silhetana*) Borer (*Indarbela quadrinotata, Xystrocera festiva and Xystrocera globosa*) Sapsuckers (*Oxyrhachis tarandus*)	Rootrot (*Ganoderma lucidum*) Heart rot (*Ganoderma applanatum*) Leaf spot (*Cercospora albizae, Phyllosticta albizzinae*) Canker (*Fusarium solani*) Rust disease (*Ravenelia sessilis*)
6	*Albizia procera*	Borer (*Xystrocera globosa*) Defoliators moth (*Pryeria sinica*)	Rootrot (*Ganoderma lucidum*) Heart rot (*Ganoderma applanatum, Polyporus gilvus*) Bark and stem disease (*Fusarium solani*)
7	*Anogeissus latifolia*	Defoliator (*Maurilia iconica, Lamida carbonifera*) Leaf skeletonising (*Orgyia postica*) Bark eater (*Indarbella quadrinotata*)	Spongy rot (*Daodalea flavida, Trametes* spp.) White Pocket rot (*Polyporus gilvus*)
8	*Anogeissus pendula*	Bark borer (*Olenecamptus anogeissi*)	White spongy rot (*Polystictus* spp.)
9	*Azadirachta indica*	*Neem scale* (*Palvinaria maxima, Aspidiotus orientalis*) Sap succer (*Helopeltis antonii*) Defoliators (*Cleora cornaria, Odites atmopa*)	Rootrot (*Ganoderma lucidum*) Powdery mildew (*Oidium* spp.) Blight disease (*Corticum salmonicolor*) Leaf spot (*Cercospora subsessilis*) Leaf spot and blight (*Pseudomonas azadirachtae*)
10	*Casuarina equisetifolia*	Attack nursery seedlings (*Brachytrupes achaetinus*) Root Borer (*Coelosterna scabrata*) Bark and wood borer (*Arbeda tetraonis*) Stem borer (*Phassus malabaricus*)	Seedling blight, stem canker and root rot (*Rhizoctonia solani*) Canker (*Phomopsis casuarinae*) Pink disease (*Corticum salmonicolor*) Black blister (*Trichosporium vesiculosum*)
11	*Dalbergia latifolia*	Bark feeder (cosmet moths) Sap feeder (*Coccus viridis*)	White rot (*Polyporus gilvus*) Spongy rot (*Trametes corrugata*)

12	*Dalbergia sissoo*	Leaf defoliator (*Plaeoptera reflexa, Dichomeris eridantis, Ascotis selenaria*) Leaf miner (*Leucoptera phemograpts*) Leaf roller (*Apoderus sissu*) Sap sucking bug (*Droscicha magifera*)	Root rot (*Ganoderma lucidum*) powdery mildew (P*hyllactinia dalbergiae)* Wilt disease (*Fusarium solani*) Root and butt rot (*Phellinus gilvus*) Leaf blight (*Rhizoctonia solani*) Rust disease (*Maravelia achroa*)
13	*Eucalyptus tereticornis*	Gall (*Leptocybe invasa*)	Die back (*cylindrocladium scoparium, cylindrocladium quinquiseptatum*) Root rot (*Ganoderma lucidum)* Pink disease (*Corticum salmonicolor*)
14	*Grevillea robusta*	Defoliator (*Looper caterpillar*)	Leaf spot or leaf blight (*Cercospora* and *Phyllostica fungi*) Root rot (*White Fungus or Candidiasis*) Leaf spot (*Phyllosticta fungus*) Wilt disease (*Cinnamon fungus*)
15	*Hopea parviflora*	Borer (*Xyleborus morastati, Massicus venustus*) Sap suckers (*Nodostoma, Rhynchites and Eups nietneri*)	Root rot (*Fomes lamaoensis*) Heart rot (*Trametes spongipellis*)
16	*Lagerstroemia microcarpa*	Leaf webbing (*Phycita* sp., *Indomias cretaceus*)	Seed rot (*Drechslera* sp., *Fusarium oxysporum,* and *Curvularia lunata*) Damping-off (*Rhizoctonia solani*) Collar rot (*Rhizoctonia solani)* Leaf spot (*Glomerella cingulata*) Mosaic disease (*L. microcarpa*)
17	*Leucaena leucocephala*	Leaf foliater and branch eater (*Aspidiotus lataniae*) Shoot and stem borer (*Xyleborus morigerus*) Seed borer (*Araecerus fasciulatus*) Defoliater (*Heliothis armigera*)	Leaf blight (*Collotorium trifolli*) Attack Stem and root (*Fusarium semitectum*)

18	*Melia dubia*	Root damage, dropping (*Odontotermes obesus*) Mould disease (*Coccid* spp.) Sap sucker (*Chorthippus curtipennis*) Defoliator (*Orlharis* spp.)	
19	*Pongamia pinata*	Leaf *miner* (*Lithocolletis virgulata* Meyrick) Defoliator (*Maruca ambionalis* Felder) Stem borer (*Leiopus nebulosus* Linnaeus) Bark borer (*Indarbela quadrinotata* Walker) Common banded awl (*Hasora chromus* Cramer) Leaf gall mite (*Aceria pongamiae* Keifer) Sap feeders (*Ferissia virgate* Cockerell) Pod borer (*Ephestia* sp.)	Leaf spot and blight (*Fusicladium pongamiae*) Leaf Rust (*Ravenelia hobsoni, Ravenelia stictica*) Powdery mildew (*Oidium* spp.)
21	*Pterocarpus marsupium*	Sapwood borer (*Xylotrechus subscutellatus*) Defoliator (*Hyblaea puera*)	Leaf sport (Glomerella cingulata) Root rot (Fusarium solani)
22	*Pterocarpus santalinus*	Defoliator (*Hyblaea puera* Cramer)	Wilt and root rot (*Fusarium solani*) Leaf spot (*Cylindrocladium scoparium*)
23	*Santalum album*	Stem borer (*Zeuzera coffees* Nietn) Bark feeder (*Indarbela quardinotata* Walker) Heartwood borer (*Octofasiculata aurivillius*)	Spike disease (*Nephotettix virescenes*)
24	*Shorea robusta*	Heartwood borer (*Hoplocerambyx spinicornis*) Sap suckers (*Drosicha stebbingii,Pedroniopsis*) Defoliators (*Lymantria mathura*) Seed and seedling pest (*Pammena theristis*)	Heart rot (*Fomes caryophyllii, Hymenochaete rubiginosa and Trametes incerta*) Butt rot (*Fomes fastuosus*) Root rot (*Polyporus shoreae*)
25	*Swietenia macrophylla*	Serious pest (*Hypsipyla robusta, Hypsipyla grandella*)	Root decay (*Fomes noxius*) White pocket rot (*Polyporus gilvus*)

26	*Tectona grandis*	Serious pest (*Pagyda salvalis, dichocrocis punctiferalis*) Stem borer (*Dihammus cervinus, Sahyadrassus malabaricus and Zeuzera coffeae*) Defoliator (*Hyblaea puera*) Teak skeletonizer (*Eutectona machaeralis*)	Leaf blight (*Rhizoctonia solani*) Wilt disease (*Pseudomonas solanacearum*) Root rots (*Armillaria mellea, Phellinus lamaoensis, Rigidoporus lignosus*) Pink disease (*Corticium salmonicolor*) Stem canker (*Nectria haematococca*) Die back (*Stemphylium spp.*) Top dying (*Phomopsis tectonae*) White rot and heart rot (*Fomes lividus, Polyporus zonalis*) Powdery mildew (*Uncinula tectonae*) Leaf rust (*Olivea tectonae*)
27	*Terminalia arjuna*	Leaf roller *(Apoderus trancuedaricus)* Defoliator *(Gelasma goniaria, Laymantria mathura)* Wood borer *(Ceranistes arboreus, Lyctus africanus and Sinoxylon anale)* White wax insect *(Ceroplastes ceroferus)*	White fibrous rot (*Polystrictss affinis*)
28	*Terminalia bellerica*	Plant feeder (*Trabala vishnou)* Defoliator (*Lamida carbonifera)* Bark borer (*Thamnurgides indicus , Thamnurgides opacifrons)* Felled timber attacker (*Aeolesthes holosericea*)	Galls disease (*Puccinea terminaliae*) Powdery *mildew* (*Phyllactinia terminaliae*) Heart rot (*Fomes fastuosus, Fomes robineae, Fomes senex* and *Ganoderma applanatum*) White spongy rot (*Trametes lactinia*)

14

To Study the Harmful Insects and Pests of Temperate Tree Species

Objective: To study the harmful insects and pests of various temperate tree species.

Observation recording: The students are advised to visit the different forests areas and identify various insects and pests damaging trees in various ways like leaf feeding (defoliator), stem boring (stem borer) damaging seeds (seed borer) etc and various pathogens. Collect the specimens from field bring them in laboratory and identify with the help of inhouse specimens in college laboratory or standard literature viz. The Ecology and Control of the Forest Insects of India and the Neighbouring Countries by Beeson CEC, Forest Pathology: Principles and Practice in Forestry by Bakshi, BK. Some examples of various temperate tree species are given in below table:

S. No.	Scientific Name	Insects	Disease and pathogen
1	*Abies pindrow*	Bark beetle (*Ips longifolia*) gal forming Aphids (*Chremes abietis piceae*)	Heart rot (*Fomes pini*) White fibrous rot (*Poria corticola*)
2	*Cedrus deodara*	White grub (*Grandia albosparsa*) root feeding insects (*Popillia cyanea, Melolontha furicauda*) Deodar looper (*Ectropis deoderae*) Bark & stem borer (*Cateremna cedrella*)	Root& but rot (*Heterobasidium annosum*) damping off (*Fusarium sp., Rhizoctonia solani*)
3	*Cupressus torulosa*		Cypress canker (*Monochaetia unicornis*)
4	*Juglans regia*	Stink bug (*Apodiphus pilpes*), green aphid (*Chromaphis juglandicola*)	Powdery mildew (*Phyllactinia quttata*) stem canker (*Cytosperma leucosperma*)

5	*Picea smithiana*	Foliage feeder nymphs (*Chermes abietis-piceae*) seed borer larvae (*Dioryctria abietteta*) young bud feeder (*Eucosma hypsidryas*)	Rust (*Peridermium piceae*) rot (*Fomes pinicola*)
6	*Pinus kesiya*	Shoot-borer (*Dioryctria castanea*)	
7	*Pinus roxburghii*	Cone feeder larvae (*Chlorophorus strobilicola*), foliage feeder larvae (*Cryptothelea crameri*), sap wood borer (*Sphenoptera atarrima*), Bettel (*Polygraphus longifolia)*	Fusarium wilt (*Fusarium oxysporum*) rust (*Cronartium himalayense*)
8	*Pinus wallichiana*	Beetle (*Hylobius angustus*)	Brown rot (*Polyporous schweinitzii*), Needle blight (*Lophodermium macci*)
9	*Quercus floribunda*	Defoliator (*Heterocrasa expansalis, Malacosma indica*), Acorn borer (*Eucosma dryocarpa*)	
10	*Quercus leucotrichophora*	Darkling beetle (*Uloma scita*), Fan-bearing Wood Borer (*Ptilinus binodulus*), Gypsy moth (*Lymantria obfuscata*)	Heart rot (*Fomes conchatus*) Root rot (*Fomes annosus*)
11	*Quercus semecarpifolia*	Fire beetle (*Melanophila coriacea*), Oriental Wood borer/Boxwood borer (*Heterobostrychus aequalis*)	
12	*Salix alba*	Beetle (*Cryptorrhynchus lapathi*) Musk beetle (*Aroma muschata*)	Canker disease (*Cytospora chrysosperma*)

15

To Study the Planting and Stand Management Practices of Teak (*Tectona grandis*)

Objective: To study the planting and stand management practices of Teak (*Tectona grandis*).

Observation recording: Students are advised to visit the teak forest or plantation for seed collection then sowing the seeds in nursery and record the various nursery practices and conduct tending operations in teak plantation. Some examples are given below:

Seed collection and storage

Seeds should be collected from 10-year-old trees. Although fertile seed is available from young tree. In a year two collection February-March and October-November

Capsules should be dries in shade for one day

Gently shaken to clean the shells

Dry and stored in a tins at cool dry places

3,67,400 seeds weigh in one kg

Nursery techniques

Seed should be sown in raised nursery bed in the month of October-November or February-March.

Soil should be sterilized with Aldrex or BHC against termites.

Seed should be sown 20 gm/m^2 of bed in lines 10 cm apart and 2.5 mm deep.

Germination takes place within 5-15 days of sowing.

60-75 seedlings are obtained from one gram of seeds.

Seedlings are pricked out when they attain a height of 4-8 cm.

Pricking of seedling should be done in to polythene bag.

Polythene bags initially should be placed in shade for few days and then shifted in the open in sunken beds.

The seedlings are hand watered for some time and later the beds may be irrigated.

Planting out

The area to be planted is clear felled ploughing and properly soil working should be done

In Terai areas of UP for the first year only cash crops are raised and plantings is done in the 2nd year.

Pit size should be 45 cm^3 or 30 cm^3

For saline and alkaline soils large pit of 60 cm^3 with soil replacement with good soil are dug out at wider spacing

Planting is done in July in areas subject to South-West Monsoon

October-November in areas of North-East Monsoon

Container/bag should be removed at the time of planting

Watering and fertilizer application should be done in poor soils

Casualties are beaten up immediately after planting in the first year

In arid areas 3-4 watering in the first year produce good stock

Spacing

Spacing depends upon the objective of plantation

Spacing varies 1.8m × 1.8m – 3m × 3m

For firewood/fuelwood: 1m × 1m or 1.5m × 1.5m

For pulpwood/poles: 2m × 2m or 3m × 2m

For saw log: 3m × 3m

For Windbreaks and shelterbelts

1.5m × 1.5m or 2m × 1m

Vegetative propagation

Clonal propagation has been attempted by means of Grafting, budding, air layering and soft wood cutting.

Propagules should be treated by IBA 10000 ppm

The selected plus tree are cut at about 12 cm height from the ground level

The coppice shoot are harvested between 45 days and 55 days after the tree has been cut

The cut shoot are placed in a bucket of water transplanted to the nursery

The length of cuttings varies from 10-15 cm

The base of the cuttings should be treated with 200 ppm solution of Benlate Fungicide for 15-30 minutes.

The base up to 2 cm is treated with 6000 ppm IBA diluted with talc powder

The cuttings are planted in polythene tubes

The growing media is vermiculite

Cutting develop good root system within 5-6 weeks

Tending operations

Tending operations such as, weeding, fertilization thinning etc. should be done time to time

16

To Study the Planting and Stand Management Practices of Kala Shisham (*Dalbergia latifolia*)

Objective: To study the planting and stand management practices of Kala Shisham (*Dalbergia latifolia*).

Observation recording: Students are advised to visit the shisham plantation or forest for seed collection then sowing the seeds in nursery and record the various nursery practices and conduct tending operations in shisham plantation. Some examples are given below:

It is one of the simplest species to proliferate through almost all the common methods *viz.* direct sowing, entire transplanting and stump planting.

Seed collection and storage

The seeds are sufficient amount in every year and keep viability for one year. Fertile seeds are produced at the age of 3-4 years, but it is advisable to collect seeds from the middle aged vigorously growing trees having straight and clean boles; on an average a medium sized tree produces 12-15 kg pods (4-5 kg clean seed).

Time of collection varies from December to mid-February in Assam; West Bengal, Punjab and from November to March in Bihar and Orissa. The seed can be collected by climbing the tree or beating off with sticks. The pods are dried in the natural light for 3-4 days and stored after removing dead leaves and foreign matter etc.

Germination capacity and plant percent

Germination capacity and plant percent in *Delbergia latifolia* are about 90 and 45 per cent respectively.

Nursery technique

Seeds soaked in water germinate after 7 to 15 days. Pricking out by end of first season (18-20 cm height). About 56 kg seeds are enough for one ha area.

Planting technique

1. Direct sowing

Direct sowing is the easiest method to raise shisham plantations; in fact, most of the seeds are sown in lines at the break of rains. Distance between lines varying from 3 to 4m.

2. Entire planting

The planting of entire seedlings being much more laborious and costly, is adopted only in particular cases such as arid areas, areas infested with tall grasses, wet sides or along the roadsides. Planting is generally done when the summer rains have properly set in, while the pits may be dug sometimes prior.

3. Stump planting

Stump planting has clear advantages over either entire planting or direct sowing. Cost of stump planting is low; planting period is extendable from July to September. The season of stump planting would depend upon the local rainfall and availability of artificial irrigation. Best time for planting is the rainy season. Stumps are spaced 1.8 m apart in lines on trenches which are 3 m apart from row to row.

Stand Management Practices

Weeding: Frequent weeding is necessary for first several years.

Pruning: Pruning of young trees helps to proper growth and development.

17

To Study the Planting and Stand Management Practices of Sandal Wood (*Santalum album*)

Objective: To study the planting and stand management practices of Sandal wood (*Santalum album*).

Observation recording: Students are advised to visit the Sandalwood plantation or forest for seed collection then sowing the seeds in nursery and record the various nursery practices and conduct tending operations in Sandal wood plantation. Some examples are given below:

Artificial regeneration: Abundant plantations dating from 1870 have been found in Mysore, Coorg, North Coimbatore and Nilgiris, partly transplanting and partly by dibbling or sowing in situ. In the majority of cases, they failed. One of the chief causes of failure was the incomplete realization of the parasitic nature of sandal. Another reason was the non-availability of shade in the early period, the sowing having been done in the open. Again, wrong tending, as by sudden reveal of the lateral covering, which, in the case of young trees, accustomed to lateral shade, results in bark scorching. It commences with cracking and peeling of the bark in thin strips and later develops in long, more or less triangular. Browsing and in attention rust lift accounted for a number of failures.

Artificial regeneration may be brought about by

1. **Direct sowings**
2. **Nursery Bed**: Sowing in beds and planting out when 2 or 3-month-old. When seedlings attain 4 to 6 leaf stage they are transplanted into polythene bags along with seeds of primary host *cajanus cajan*.
3. **Stump planting:** Upto 3 years – 2¼ to 3 years old Sandal stumps are better than younger ones from the point of view of storage stumps stored in tins with tight lids have spouted even when stored up to 20 days. Stump – planting has shown as much as 99% success.

4. **Planting pieces of roots**
5. **Root suckers**

   Direct sowings cheap and best way – sown in April. Two recognized methods of procedure (1) sow over a large area at very wide and irregular spacements. Supervise and control difficult. A better method is to concentrate sowing on area with alternate to protection and trust to birds to spread the seeds from trees later. In both the methods make numerous hole ½” deep with pointed stock without disturbing the soil unnecessarily and put one seed in each.

Localities to be regenerated: (1) Open blanks – Hosts to be established in advances, (2) Shrubby jungle – Together with hosts.

Hosts are suspected to be highly significant in conferring relative resistance, against spike. *Eg., Strychnos nuxvomica, Pongamia glabra, Azadirachta indica,Cajanus cajan* and *Cassia siamea* are stated to give high resistance. Lantana is good but is suspected to harbour an insect vector.

Tending in earlier stages

1. Maintenance of superior host plants and their tending
2. Provision for tree and natural expansion of the crown
3. Feeing of suppressed plants
4. Providing light lateral shade to avoid sun-scorch

In the later stages

1. Tending of hosts plant and
2. Weeding, Cleaning and Climber cutting

18

To Study the Planting and Stand Management Practices of Poplar (*Populus deltoides*)

Objective: To study the planting and stand management practices of Poplar (*Populus deltoides*)

Observation recording: Poplar rarely raised by seeds. It is generally raised by stem cuttings. Students are advised to visit the Poplar plantation for collection of stem cuttings then raised in polythene bags and record the various nursery practices and conduct tending operations in Poplar plantation. Some examples are given below:

Artificial Regeneration: Poplars can be raising quality planting materials (QPM) by the following methods.

i) By Sexual Reproduction

Poplar seeds are very minute and 14,000 seeds weigh one kg. The seed of *Populus deltoides* has been germinated under laboratory conditions with germination percentage varying from 5.25 to 19.25%. The medium of germination, *i.e.* two parts soil and one part of river sand (2:1 ratio) duly sterilized proved most effective and gave higher survival and best shoot development. Germination tests and viability conducted of the seed after 6 years of storage showed that-20°C storage was superior to -50°C storage in maintaining seed viability.

The seedlings are rarely raised by seeds.

ii) Vegetative propagation

1. **Cuttings:** Take the poplar Cuttings 18-25 cm long and 1-2.5 cm thick having at least three to four buds are directly planted in the field in well worked soils, keeping one bud above the ground. The soil around the cuttings is compacted.
2. **Sets:** A set is one year old cut back plant without root or a long stem cutting which consists of one year growth. The root portion left in the nursery can

again be used to produce plant/set/cuttings for the next planting season. Sets are planted in 75 cm to 1 m deep pits. The length of set is about 3.5 m or more.

3. **Bag plants:** These can be raised in polythene bags of 15cm x 23cm x 150 gauges filled with soil mix and treated with 5% aldrex. Cuttings are planted in these bags in February in the year of planting. Vigorous plants are not obtainable by this method.
4. **Entire transplants (ETPs):** This is the best method for transplanting poplars. These are one- or two-year-old plants raised from cuttings in the nursery. While removing these plants, the root is cut at 25 cm depth and all side roots more than 10 cm long are also trimmed. The plants are taken out with naked root and without foliage.

Nursery Technology

i) Site Selection

The two important considerations for the selection of site for nursery are soil type and irrigation facility.

Poplar (*Populus deltoides*) needs prefer sandy loam, deep, fertile, well drained, neither alkaline nor too acidic, free from underlying impervious layer and rich in organic matter. Soils with top-layer removed by erosion and site with laterite occurring within 60 cm of surface are considered unfit for poplar cultivation.

Assured irrigation is essential requirements of a good nursery site. As poplars are very sensitive to zinc deficiency, zinc sulphate @ 25 kg/ha should be applied in the area at the time of site preparation.

Site Preparation

A light preparatory irrigation should be given so that the soil has proper moisture at the time of Ploughing. The field should be levelled after ploughing. One deep ploughing upto 25-30 cm depth is desirable. Phosphatic and Potassium fertilizers should be added at this point. After addition of fertilizers, one more harrowing should be arranged.

Clones

A lot of good quality clones have been developed and launched commercially. Some of them are G-48, S7C15, S7C8, L-34, L-49, L-52, Uday, Kranti, WS-39, WSL-22, WSL-27, WSL-32, WSL A-49, and PL-1 to PL-7 *etc.*

ii) Selection of cuttings

Cuttings from the main stem of the poplar give better results than those obtained from branches. One year old trees from existing nurseries are used for preparation of cuttings for the new nurseries. Diseased, dying off-type and suppressed plants must be remove from the beds where stock is reserved for preparation of cuttings for the new nurseries. The plants can be cut 2 cms from ground level and the resultant sets should be graded and formed into bundled and should be transported to the new nursery site immediately and stored there in the storage pits, which should be kept full with fresh water to be replenished every day.Cuttings made from ripened one year old wood of the main stem give better results. Cuttings obtained from young, healthy and vigorous trees perform better. The length of cuttings should be 18-25 cm and these should contain three to four buds. When cuttings are taken from the trees, the age of trees and of the parts of trees from which the cuttings are taken markedly affect rooting (Zsuffa, 1976).

Preparation of cuttings

Cuttings should be prepared with a very sharp and fairly heavy tool to obtain a very clean and smooth cut. During the preparation of cuttings, the cut must be given at a point of the set which is fully supported on a log/beam of wood to prevent splitting of cuttings. Maximum number of available cuttings from each plant down to one centimeter diameter cans be used for planting. All cuttings must be submerged under fresh water in drums immediately after preparation of the cuttings.

Treatment of cuttings

Before planting, cuttings are soaked in water for 24 hours. Cuttings should be treated with Aldrin (250 ml of Aldex 30 EC in 100 litres of water) for 10 minutes each and then only planted in the nursery.

iii) Planting of cuttings

For production of healthy and vigorous entire plants, planting of cuttings at 80 cm x 60 cm is the best spacing.To produce one year old plants, spacing of 1-1.4 m between the lines and of not less than 40 cm between the plants is normally adopted. The best time of planting cuttings in the nursery is January to February. Each cutting with its thinner end up should be planted in the hole in such a way that the upper portion is just 2 cm above the ground level.

Cultural Practices

Irrigation in Nurseries

Irrigation should be provided as soon as the planting of cuttings in any bed is completed. The first irrigation should be medium heavy so that about 5 to 7 cm water are uniformly above ground level at the time of irrigation. Soil moisture in the nursery bed should be kept high during rooting of cuttings. Consequent irrigation should be light and the interval may vary between 7 to 10 days depending upon the type of soil. Light sandy soils will need frequent irrigations whereas medium to heavy soils will need irrigation at longer intervals. The top soil should not be allowed to develop cracks and become absolutely dry. After the rainy season one to two irrigations per month is adequate.

Fertilizer application

Poplar tree grow very fast, the nursery soil has to be enriched frequently. Urea, Super phosphate, murate of potash and plentiful supply of Farm Yard Manure are essential for maintaining the growth of cuttings. Application of well decomposed compost or farm yard manure @ 200 to 250 quintal per hectare nursery area at the time of soil working is very helpful for the growth of poplar.

After the rains have set in, 2g of urea per plant is given; Singling of collar shoots should be done during April-May when the most vigorous shoot has attained a height 30-35 cm. Adequate care should be taken to ensure that the young leaves are not damaged.

19

To Study Planting and Stand Management of Bastard Mahogany (*Swietenia macrophylla*)

Objective: To study planting and stand management of Bastard Mahogany (*Swietenia macrophylla*)

Observation recording: Students are advised to visit the Mahogany plantation for collection of seeds/stem cuttings then raised in polythene bags and record the various nursery practices and conduct tending operations in Mahogany plantation. Some examples are given below:

Seed sowing

According to established practices, Swietenia macrophylla, commonly known as mahogany, can be propagated using seeds. For optimal results, it is recommended to select seeds from a healthy and well-maintained mother tree. The seeds should be planted in furrows or holes that are approximately 3-7 cm deep, within a light sand bed. To facilitate germination, it is advisable to keep the seeds in shaded areas and ensure they remain adequately moist. When planting Mahogani, it is generally recommended to maintain a spacing of 2-3 meters between individual plants.

Weeding

In order to promote optimal growth and survival of seedlings, it is necessary to conduct regular weeding in Mahogani plantations.

Fertilizing

According to research and established practices, altering the chemical composition of the apical shoots and promoting vigor through fertilization can potentially reduce the susceptibility of Mahogany seedlings to shoot borer damage and enhance their tolerance. After planting, it is common practice to apply fertilizer in a ring around the seedlings, typically at a rate of 75-100 grams of NPK (chemical fertilizer) per plant. During the rotation period, gap filling through replanting is often performed twice. The first replanting is typically conducted one month after the initial planting to replace any dead seedlings, while the second replanting occurs at the end of the second year.

Pruning

Pruning is an essential practice aimed at removing dead or non-productive branches from the lower trunk to stimulate the growth of clear wood in trees. It also helps to reduce the risk of diseases and pest infestations such as the shoot borer. Typically, pruning is recommended during the initial three years of tree growth, as this period is crucial in minimizing the vulnerability to shoot borer attacks, as the pest predominantly targets young trees. The ideal time for pruning is shortly before the onset of the rainy season.

Thinning

Thinning is primarily undertaken to enhance the growth of the remaining trees, ensuring they develop a desirable form for the final crop. During thinning, careful consideration should be given to removing trees affected by disease or pests, as well as those that are deformed, poorly shaped, or suppressed in growth. The selective removal of diseased trees can contribute to the genetic improvement of the final crop, making it less susceptible or more tolerant to potential attacks.

Control of pests and diseases

The shoot-borer *Hypsipyla robusta* poses a significant threat to Mahogani plantations, causing severe damage. This pest primarily targets the terminal shoots of saplings and pole-sized trees, resulting in dieback symptoms and the deformation of affected trees. The larvae bore into the budding shoots of saplings, leading to damage to the terminal bud, stem forking, and stunted growth. Currently, there is no known effective method to control this insect. However, preliminary studies have shown that heavy trimming during the first three years following planting can help reduce shoot borer infestations. Additionally, interplanting *Acacia mangium* around mahogany plantations has shown promising results in averting *H. robusta* infestations. Similarly, interplanting neem (*Azadirachta indica*) with Mahogani has been observed to reduce shoot borer infestations.

20

To Study Planting and Stand Management of Khair (*Acacia catechu*)

Objective: To study planting and stand management of Khair (*Acacia catechu*)

Observation recording: Students are advised to visit the Khair plantation for collection of stem cuttings then raised in polythene bags and record the various nursery practices and conduct tending operations in Khair plantation. Some examples are given below:

Site selection: When selecting a site for planting Acacia catechu, it is recommended to choose an area that receives abundant sunlight and has well-drained soil. Ideally, sandy or loamy soils with a pH ranging from 6 to 8 are preferred for this tree species. It is important to avoid planting in locations with a history of water logging or where the water table is high.

Planting: For planting Acacia catechu, it is recommended to choose the rainy season, particularly during the months of June or July. Select seedlings that are 6-12 months old, with a stem diameter of at least 1 cm. Plant them at a spacing of either 3 m x 3 m or 4 m x 4 m. There are three primary methods for propagating Acacia catechu: (1) direct sowing, (2) planting nursery-raised seedlings, and (3) stump planting. Direct sowing is generally preferred over stump planting, while planting out entire plants without containers is not as desirable.

i) Line/strip sowing in lines 60 to 90 cm apart, strips 2.5 to 3 cm apart.

ii) Broadcast sowing in weed free area.

iii) Patch sowing by dibbling 2.5 to 3 cm apart.

iv) Potted plants or with ball of earth.

Planting technique

a) Polybag plants

To prepare for planting, a pit with a volume of approximately 30 cm^3 should be dug, ensuring a spacing of 3 m × 3 m between the pits. It is recommended to prepare the pits approximately two months in advance of the planting date. Planting is typically conducted in the month of July. When planting, it is important to remove the polythene container from the seedling before placing it in the prepared pit.

b) Ball of the earth planting

During the uprooting of the seedling, it is common practice to cut the taproot at a depth of 25-30 cm. However, it is advisable to avoid splitting or bruising the taproot during this process. Transplanting of the seedling is typically carried out during the rainy season, as winter transplanting tends to result in failure.

c) Stump planting

Stumps for planting are typically prepared from 12-15-month-old seedlings. The recommended dimensions for the stump are a root portion of 23-31cm and a shoot portion of 2.5 to 5.0 cm. Stumps with a collar diameter less than 1 cm tend to have poor survival rates. Stumps can be stored for up to 3 days before planting. It is advisable to plant the stumps at the onset of the monsoon season.

Silviculture system for management

According to research and established practices, the clear felling followed by artificial regeneration is considered the most effective management system for Khair (Acacia catechu) trees.

Disease

Ganoderma lucidum causes root rot disease with considerable mortality in khair plantation raised after clear felling.

Weed control: To ensure optimal growth of the plantation, it is crucial to maintain weed-free conditions during the initial two years after planting. Weeds can be effectively managed through manual or mechanical methods, including hand weeding or the use of appropriate tools to remove unwanted vegetation.

Fertilization: For effective fertilization of the plantation, it is recommended to apply nitrogenous fertilizers such as urea at a rate of 50-60 kg/ha/year. Phosphorus fertilizers like single superphosphate should be applied at a rate of 100-150 kg/ha/year, while potassium fertilizers such as muriate of potash should be applied at a rate of 50-60 kg/ha/year.

Pruning: Prune the trees in the third year after planting to promote the growth of lateral branches. Remove any dead or diseased branches.

Thinning: Thinning of the plantation is recommended when the trees reach the age of 8-10 years. This process involves removing poorly formed or diseased trees to promote better growth and overall health of the remaining trees. It is advisable to maintain a spacing of 5-6 m between the trees after thinning.

Harvesting: The harvesting of trees is typically conducted when they reach the age of 10-15 years, depending on the specific requirements for timber size and quality. The trees are cut at ground level during harvesting, and the bark is removed. The bark holds medicinal properties and can be utilized for various purposes.

21

To Study Planting and Stand Management of Safeda (*Eucalyptus tereticornis*)

Objective: To study planting and stand management of Safeda (*Eucalyptus tereticornis*)

Observation recording: Students are advised to visit the Safeda plantation for collection of stem cuttings then raised in polythene bags and record the various nursery practices and conduct tending operations in Mahogany plantation. Some examples are given below:

Eucalyptus tereticornis is a fast-growing species of eucalyptus tree that is widely planted for its timber, pulp, and biomass production.

Site selection: Choose a site that receives plenty of sunlight and has well-drained soil. It prefers sandy loam soils having pH value 5.5 to 6.5. Avoid planting in areas with a history of waterlogging or where the water table is high.

Planting: Plant seedlings during the rainy season, preferably in June or July. Use seedlings that are 6-12 months old, with a stem diameter of at least 1 cm. Plant to plant safeda seedlings at the spacing of 3 m x 3 m or 4 m x 4 m.

Weed control: Keep the plantation free of weeds for the first two years after planting. Use manual or mechanical methods to remove weeds.

Fertilization: Apply nitrogenous fertilizers (e.g. urea) at the rate of 150-200 kg/ha/year, phosphorus fertilizers (e.g. single superphosphate) at the rate of 100-150 kg/ha/year, and potassium fertilizers (e.g. murate of potash) at the rate of 50-60 kg/ha/year.

Pruning: Prune the trees in the third year after planting to promote the growth of lateral branches. Remove any dead or diseased branches.

Thinning: Thin the plantation when the trees are 5-7 years old to remove poorly formed or diseased trees. Leave a spacing of 5-6 m between trees.

Harvesting: Harvest the trees when they are 8-10 years old, depending on the desired size and quality of the timber. Cut the trees at ground level and remove the bark.

22

To Study the Species Suitability for Various Problematic Areas

Objective: To study the species suitability for various problematic areas.

Observation recording: To know the suitability of the tree species for plantation of various problematic areas, students are advised to visit the various nearby problematic sites or forest and record the naturally growing tree species. Here, some examples of species suitability are given below:

Tree species suitable for mined out areas: *Dalbergia sissoo* (Shisam), *Albizia lebbek* (Siris), *Sterculia urens* (Kadaya), *Lagerstroemia parviflora* (Senja, Landiya), *Acacia catechu* (Khair), *Grevillea robusta* (Silver oak), *Cassia siamea* (Kasod), *Pinus roxburghii* (Chir), *Pinus caribaea*, *Terminalia arjuna* (Arjun), *Leucaena leucocephala* (Subabool), *Ficus benghalensis* (Bargad), *Anthocephalus cadamba* (Kadam), *Adina cordifolia* (Haldu), *Ixora coccinea*, *Ixora pavetta* (Lokhandi), *Syzygium cumini* (Jamun), *Ficus religiosa* (Pepal), *Acacia senegal* (Kumath), *Maytenus emarginata* (Kankera), *Eucalyptus citridora* (Safeda), *Acacia auriculiformis* (Australian babool), *Bombex ceiba* (Semal), *Mallingtonia hortensis* (Akash neem) etc.

Tree species suitable for Water logged

Waterlogged areas are created due to excessive and faulty canal irrigation, high water table, high rainfall but poor drainage or poor infiltration due to underneath hard Kankarpan formation. These areas can be categories in two groups

a) ***Fresh Water species***: *Syzygium cumini*, *Terminalia arjuna*, *Eucalyptus robusta* (Swamp mahogany), *Tamarix dioica* (Jhau), *Barringtonia acutangula* (Samudrafal), *Vachellia nilotica* (desi babool), *Mallotus nudiflorous* (Pindar), *Acacia auriculiformis*, *Bischofia javanica* (Paniala), *Trewia nudiflora* (Pindar).

b) ***Marine water species***: *Avicinea officinalis* (Bina), *A. marina* (Tivar), *Ceriops decandra* (Kirari), *Rhizophora apiculata* (Kantal), *Kandelia candel* (Kandal), Sonneratia alba (Pandhari chipi) etc.

Tree species suitable for rocky barrens

Rocky barrens occur in the form of hillocks due to heavy soil erosion. The various shrubs like *Grewia tenax* (Gangan), *Rhus mysorensis* (Dansar), *Helecteris isora* (Marodfali), *Dodonaea viscosa* (Sannata) and tree species such as Capparis grandis (Indoka), *Sterculia urens* (Kadaya), *Lannea coromandelica* (Jhigan), *Butea monosperma* etc.

Tree species suitable for saline and alkaline soils

Saline soils contain excessive soluble salts of sodium, magnesium, potassium, chloride, sulphate, bicarbonate etc. Alkali soils have large concentration of sodium carbonate type salts, low permeability, pH8.5-10.6 and ESP greater than 15. These soils are known as *usar* in Uttar Pradesh. Plant species like *Tamarix articulata, Atriplex spp., Salvadora oleoides, S. persica, Acacia nilotica, Capparis decidua, A. leucophloea, Butea monosperma, Ziziphus mauritiana* etc.

Tree species suitable for ravine lands

Due to severe erosion ravines are created and these are characterize by hillocks and ridges. Tree species like *Albizzia lebbeck, Acacia catechu, Prosopis juliflora, Dalbergia sissoo, Terminalia arjuna, Azadirachta indica, Gmelina arborea, Syzygium cumini, Mitragyna parviflora* etc.

Eroded and Denuded hillslopes

The soils of eroded and denuded hillslopes are stony, shallow and deficit in moisture as well as nutrients. Plant species like *Robinia pseudocasia, Gmelina arborea, Pinus roxburghii, Pinus cassia, Celtis australis, Dendrocalamus strictus, Acacia catechu, Tectona grandis, Vitex negundo, Dodonea viscosa* etc.

Mined out areas

The mining is done to extract ore, coal, zinc, copper and other materials. Tree species like *Dalbergia sissoo, Acacia catechu, A. auriculiformis, Holoptelia integrifolia, Pinus caribaea, Senna siamea, Albizzia lebbeck, Lannea coromandelica, Sterculia urens, Parkinsonia aculeata, Pongamia pinnata* etc.

Shifting sand dunes

Sand dunes are found in Rajasthan, Gujarat, few Haryana and in Punjab. Tree species such as *Prosopis cineraria, P. juliflora, Acacia tortilis, A. senegal,*

A. catechu, *Zizyphus mauritiana*, *Tecomella undulata*, *Azadirachta indica* etc. Shrubs like *A. jackmontii* (babal*)*, *Callygonum polygonoides* (fog), *Z. nummularia*, *Leptadenia pyrotechnica*, *Lycium barbarum* etc. and grasses *Cenchrus ciliaris*, *Dicanthium annulatum*, *Lasiurus sindicus* are the suitable plant species for afforestation in Indian hot desert for establishing the marching sand dunes.

23

To Study the Composition of a Forest Area

Objective: To study the composition of a forest area.

Observation recording Lay the quadrats with the help of rope. Identify all the plant species, and count their number in each quadrat, dbh and record the observation in a tabular form as given below:

S. No	Name of plant species	No. of Quadrat					Relative frequency (Rf)	Relative density (Rd)	Relative dominance (Rdo)	IVI (Rf + Rd + Rdo)
		1	2	3	4	5				
1	*Buchanania cochinchinensis*	5	3			8	27.27	51.61	73.78	152.66
2	*Diospyros melanoxylon*	1	3	4			27.27	25.81	18.44	71.52
3	*Madhuca longifolia*				1		9.09	3.23	0.29	12.60
4	*Adina cordifola*		1				9.09	3.23	0.29	12.60
5	*Lagerstroemia parviflora*	1		2		2	27.27	16.13	7.20	50.61

Calculation

Importance value index (IVI) is a measure of dominance and ecological success of a species. It takes into consideration Relative frequency; Relative density and Relative dominance there are following calculation:

Relative frequency: Total basal area of the species / Total basal area of all the species × 100

Relative density: No. of individuals of the species / No. of individuals of all the species × 100

Relative dominance: No. of occurrences of the species / No. of occurrences of all the species×100

24

To Study the Tending Operation Carried out in a Forest Crop

Objective: To study the tending operation carried out in a forest crop.

Observation recording: Students are advised to visit the nursey and forest plantation to record the various tending operations which are adopted by the forest department or Plantation Company. A short description about various types of tending operations is given below:

For the establishment and subsequent development of the forest crop until harvesting, various operations are conducted. These operations are implemented at different stages of the tree crop's growth to create a conducive environment for its healthy development. They are commonly referred to as tending opration, which includes:

i) Cleaning

ii) Thinning

iii) Improvement Felling

iv) Pruning

v) Climber Cutting

Weeding

During the nursery and seedling stages of a forest crop, a tending operation is performed to reduce below-ground competition for nutrients and moisture. Weeding is a crucial aspect of this operation, and various methods can be employed, like:

1. Cultural: adopting techniques like crop rotation, maintaining soil fertility etc.
2. Mechanical: by using equipment like: spade, Khurpi, hand fork etc.)
3. Biologically: by using biological entity like insect, plant, rodents etc.
4. Chemical: by applying some chemical like 2,4-D; 2,4,5-T, Paraquat etc.

Cleaning: In the sapling stage, cleaning operations are conducted to reduce above-ground competition, particularly light competition, and promote the optimal growth of desired tree species. The objective is to eliminate undesirable individuals of the same age, allowing the favored species to thrive. For instance, in plantations of *Tectona grandis* (teak), *Macaranga*, *Hopea odorata*, *Shorea robusta* (sal), *Cedrus deodara* (Deodar), and other species, cleaning involves the removal of unwanted vegetation such as Lantana, Mallotus, oak, and other competing plants.

Additionally, cleaning operations may include singling out multiple stems of important timber tree species such as teak, sal, Kardhai, and others. This process helps select the strongest and most promising individual stems for further growth, enabling them to develop into high-quality trees.

By performing cleaning activities, the growth potential of the desired tree species is enhanced by reducing competition for light and resources, leading to healthier and more productive forests or plantations.

Thinning: Thinning is a silvicultural practice carried out in an immature stand to enhance the growth and form of the remaining trees while preserving the canopy structure. The principles of thinning are specifically designed for application in pure even-aged or relatively even-aged stands, or groups of trees within a stand. These principles have been derived from the natural development of stands.

Thinning replicates the natural process of the "survival of the fittest" within a densely stocked forest. In nature, weaker or less competitive trees naturally perish, allowing the stronger ones to flourish. By implementing thinning techniques, foresters can achieve several objectives, including:

Enhanced Growth: By reducing competition for resources like sunlight, water, and nutrients, the remaining trees experience accelerated growth and development.

Improved Form: Thinning helps to maintain a well-spaced and evenly distributed canopy, leading to straighter, healthier tree stems.

Biodiversity and Habitat: Proper thinning can create openings in the forest, promoting biodiversity and providing habitats for various wildlife species.

Sustainable Harvest: Thinning can also serve as a sustainable harvest practice, providing timber while ensuring the long-term health and productivity of the forest.

In summary, thinning is a carefully planned and executed technique that mimics natural processes to foster a healthier and more resilient forest ecosystem.

Following Methods of Thinning can be adopted:

a) Mechanical Thinning, b) Ordinary Thinning, c) Crown Thinning, d) Free Thinning,

a) Mechanical Thinning

This type of thinning is typically implemented in young crops or plantations, particularly before the crown differentiation process occurs. In the case of teak plantations in Kerala, the first and second thinning operations are usually conducted at the ages of 4 and 8 years, respectively. These thinning activities primarily involve mechanical methods and entail the removal of alternate diagonal lines or rows of trees.

The objective of thinning is to reduce the stocking density of the plantation. For instance, after the first mechanical thinning, the stocking density is decreased from 2,500 trees per hectare to 1,250 trees per hectare. Subsequently, the second thinning further reduces the stocking density to 625 trees per hectare. The spacing used during thinning operations is typically 2 x 2 meters, which corresponds to approximately 2.85 x 2.85 meters. Finally, after the second thinning, the spacing is adjusted to 4 x 4 meters. These thinning practices aim to optimize the growth and development of the teak plantation by providing adequate space and resources for the remaining trees. By strategically reducing the tree density, thinning promotes healthier tree growth, improved form, and sustainable timber production.

b) Ordinary Thinning

This thinning method is known as "Low Thinning." In ordinary thinning, trees from the lower crown classes are selectively removed. The concept behind ordinary thinning is to mimic natural processes by eliminating trees that have been less successful in the competition for resources.

Ordinary thinning is widely used in forestry and is particularly suitable for light-demanding species. It is most effective when there is a market for small-sized timber in the specific species and area. The method is relatively simple to implement, involving the removal of lower crown classes of trees. It is suitable for areas where there is no risk of soil deterioration, forests infested with climbers, or where there is a potential danger of crown fireOrdinary thinning has several commands

Grades of Ordinary Thinning

i) Light Thinning (A Grade): This is limited to the removal of dead, dying, diseased and suppressed trees.

ii) Moderate Thinning (B Grade): This consists in the further removal of defective eliminated stems and whips.

iii) Heavy Thinning (C Grade): This consists in the further removal of the remaining dominated stems and such of the defective codominants as can be removed without making lasting gaps in the canopy.

iv) Very Heavy Thinning (D Grade): The distinguishing features of this grade is that, it also takes some of the dominate.

v) Extremely Heavy Thinning (E Grade): This is the heaviest thinning that can do in a crop without making permanent gaps in the canopy. More of the dominant stems even of class (a) are removed.

c) Crown Thinning

Less promising trees are removed during thinning operations to promote the growth of the best individuals. Dominated and suppressed stems are generally retained unless they are dead, drying, or diseased. This selective approach provides sufficient space and resources for the remaining trees to thrive, improving the overall health and vigor of the stand. Prioritizing the long-term success and sustainability of the forest ecosystem is essential when making such decisions during thinning operations.

d) Free thinning

Free thinning is the combination of low and crown thinning leaving properly spaced trees in upper canopy. This selective approach ensures that the remaining trees have ample space and access to resources, resulting in improved overall health and vitality of the stand.

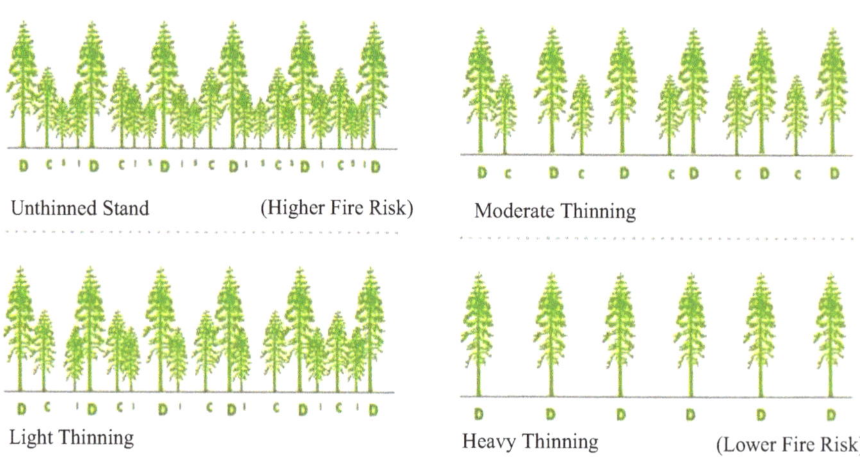

a) **Improvement felling:** Improvement felling refers to the selective removal of less valuable trees within a forest stand, aimed at promoting the growth of more valuable individuals. This practice is commonly employed in mixed and uneven-aged forests. Improvement felling may involve thinning densely stocked groups of trees, as well as clearing and providing general assistance to the young growth of valuable species. The objective is to enhance the overall quality and value of the forest stand by prioritizing the growth and development of the more valuable tree species.

b) **Pruning:** Pruning involves the removal of branches to achieve trees with clean boles. The natural elimination of branches due to environmental factors is known as natural pruning, while the removal of branches using mechanical means is referred to as artificial pruning.

Natural pruning occurs through two processes: the killing of branches and the shedding of dead branches. Lower branches tend to die when the crowns of adjacent trees are close together. This death occurs below the point of contact. Subsequently, shading of branches occurs as dead branches are attacked by saprophytic fungi, insects, and other factors, eventually breaking away due to wind or other forces. The rate of natural pruning can be influenced by manipulating the density and composition of the tree population, and it varies depending on species and habitat factors.

Artificial pruning involves the removal of moribund (dying) and lower green branches as close to the main stem as possible, typically using a handsaw. In certain cases, such as poplar and mulberry trees, artificial pruning on a large scale is not yet widely adopted in India. The height up to which artificial pruning should be conducted depends on the species, age of the tree, and local conditions. In young plantations, pruning is generally carried out up to one-half to three-fifths of the total height of the tree. The objective of pruning is to obtain knot-free timber, which can be achieved more effectively by removing lateral buds. Overall, pruning is aimed at creating trees with clean, knot-free boles, and the method employed (natural or artificial) depends on environmental factors, tree species, and management objectives.

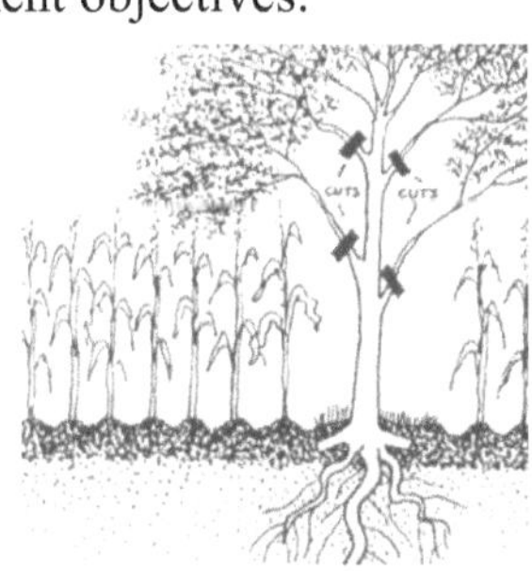

c) **Climber cutting:** It should be incorporated as part of tending operations and when marking trees for felling in the forest. Climbers pose a significant threat, especially to young trees. In older trees, they can restrict the stem's growth and degrade the quality of the wood. When cutting climbers, it is important to remove them at the base and approximately one meter above to ensure proper removal. This ensures that the climber has been effectively cut. An alternative and more effective approach for controlling climbers is to dig up their tubers during the rainy season. Ideally, the cutting of climbers should be carried out during the monsoon period. This timing enhances the effectiveness of the cutting operation.

Managing climbers is crucial for the overall health and well-being of the forest stand, particularly in ensuring the proper growth and development of the desired tree species. By actively addressing the issue of climbers through cutting and tuber removal, forest managers can mitigate the negative impacts and promote the success of the forest ecosystem.

References

Bakshi, B.K. (1976) Forest pathology: principles and Practices in Forestry, FRI Press Dehradun, Pp 400.

Beeson, C.F.C. (1941). The Ecology and Control of the Forest Insects of India and the Neighbourg Countries. Dehra Dun: Vasant Press, ii + 1007.

https://doi.org/10.1007/s00468-021-02243-3

https://indiabiodiversity.org/species/show/229714

https://indiabiodiversity.org/species/show/31727

Mishra, R. (1968) Ecology Work Book. Oxford and IBH Publishing Co., Calcutta.

Singh, K., Kaur, J. and Kumar, S. (2022) Textbook of Dendrology, Scientific Publisher, Jodhpur Rajasthan, India, Pp 173.

Troup, R.S. (1921). Silviculture of Indian Trees, Vol. 1-4, Revised and Enlarged Edition, Forest Research Institute and Colleges, Dehra Dun, 1975.

FSC
www.fsc.org
MIX
Papier | Fördert gute Waldnutzung
FSC® C083411

Zeitfracht Medien GmbH
Ferdinand-Jühlke-Straße 7
99095 Erfurt, Deutschland
produktsicherheit@kolibri360.de